Dark Wonder

Learning About Melanin

by Stephanie Abena Kaashe

c 2014

Dedication

This book is dedicated to the Ladies and Lords of Melanin whose research
and scholarship helped to inspire my passion to share this work:
Dr. Frances Cress Welsing, Dr. Ann Brown, Dr. Jewel Pookrum, Suzar,
Deborah Maat Moore, Dr. Richard E. King,
Dr. Llaila Africa and Dr. Kaba Kamene.

And to all the beautiful children of the Sun who have been graced
with the divine gift of Melanin.

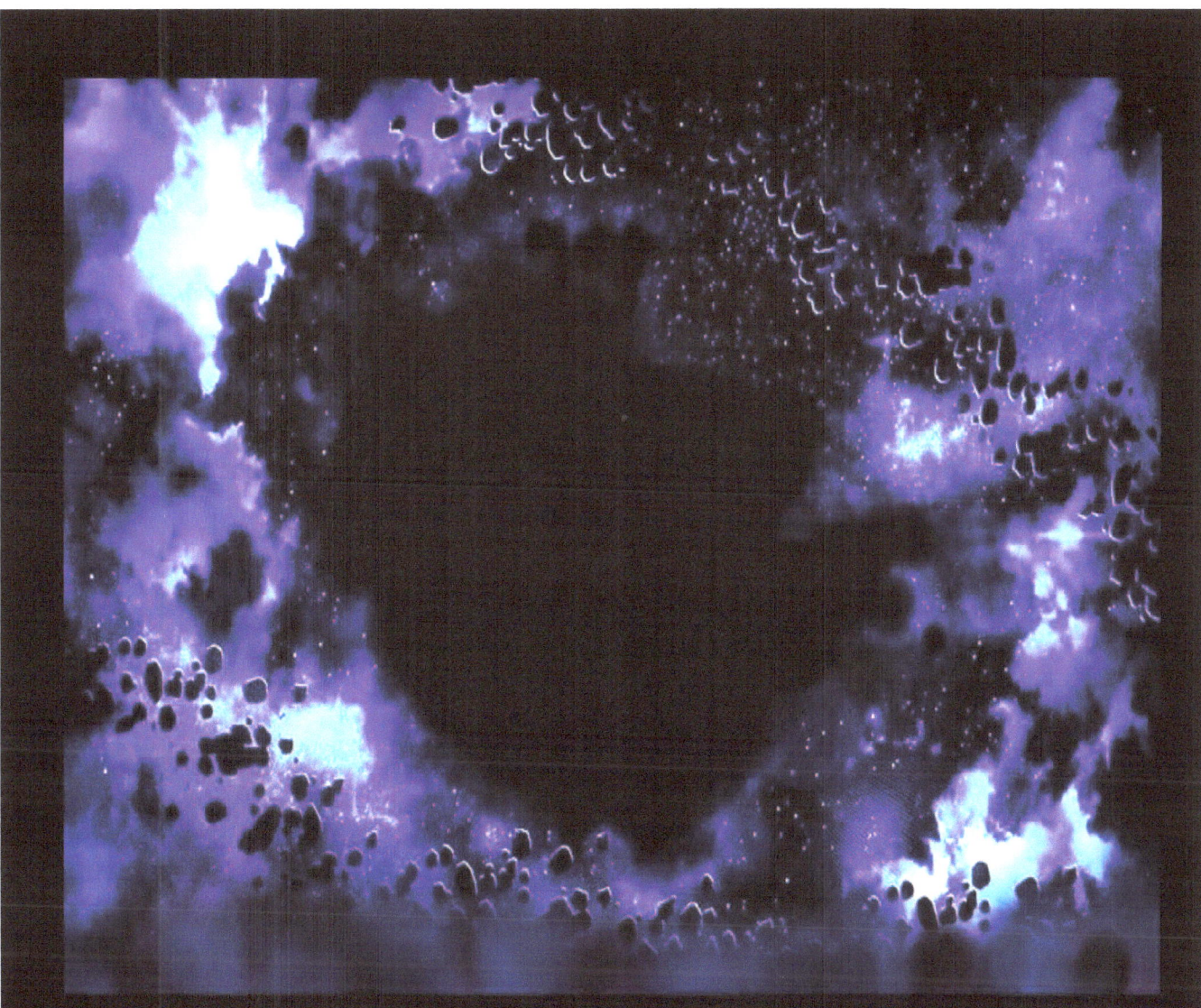

I am the substance come out of the dark

I cause pressure in space
creating a spark

3

My strength is compressed

throughout space I scatter

4

5

Then stars are born from the melanin matter

6

7

I live up above
the same as below
to serve and protect
all the life forms I know

8

I filter the sun preventing ray damage
in order to preserve the human race canvas

9

I exist on the earth in all living things
just use your third eye to see what I mean

11

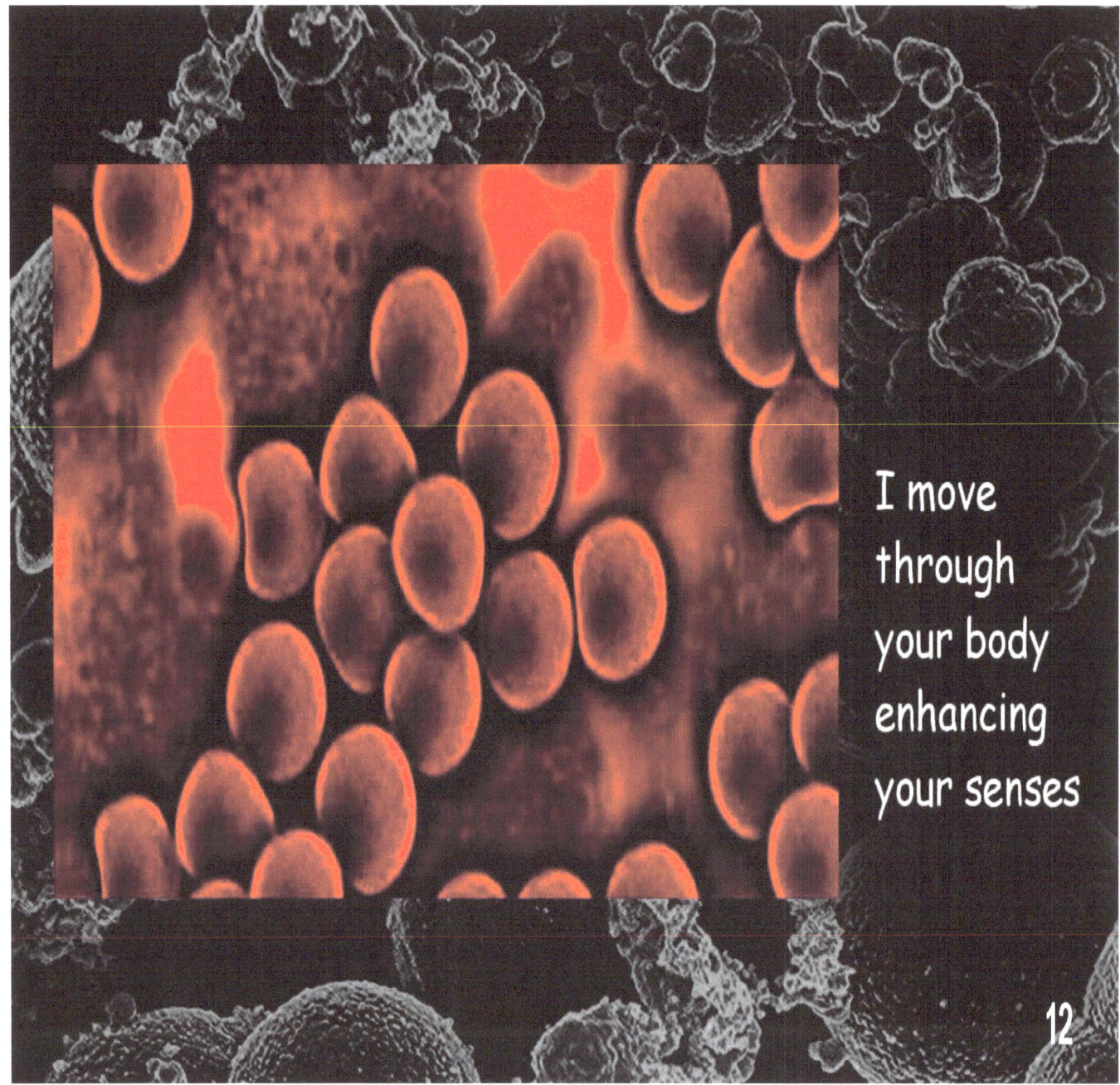

I move
through
your body
enhancing
your senses

12

When you activate me possibilities are endless

Go to bed before midnight   Eat greens   And good water

All the things
that help me
inside you
keep order

Indeed
get to know me
as much as you can

The super dark power of Melanin

What Is Melanin?

The word Melanin comes from the Greek word melanos which means black. One of the oldest African terms for black is the Kemetic or Egyptian word kem. Ancient Africans studied blackness or melanin ages ago and have provided a lot of information that is still being studied today. After several hundred years of researching the mysteries of Melanin, western scientists finally agree that Melanin is indeed the chemical key to life and living. Melanin is a vital link that connects all of creation. This carbon based substance is found in outer space among the planets and stars and throughout nature on the planet earth. Melanin is biological living light. It is an effective absorber of light and heat from the sun and is capable of changing energy into various other forms of energy. Melanin is found in nearly every organ in the human body and helps the nervous system and brain to operate. The presence of Melanin determines pigment, it is known to slow down the aging process and promote effective activity in living organisms. Melanin is the substance that gives the skin and hair its natural color as well as the iris of the eye, scales and feathers. In human beings, those with higher amounts of Melanin have darker skin color and those with less melanin show little to no pigment. Melanin plays a major role in the brain, eyes, ear, within the cells, the central nervous and circulatory systems.

Where does Melanin come from?

Melanin comes from melatonin created by the Pineal Gland (third eye or eye of Heru) in the brain. Melatonin puts Melanin into the blood stream during the night while you sleep. The body starts to make melatonin when the sun goes down and reaches a high point at midnight. After midnight, the amount of melatonin created becomes less and less until sunrise when the body starts to produce serotonin. If the body does not produce melatonin during the appointed time a person is likely to become depressed and vulnerable to disease due to a weakened immune system.

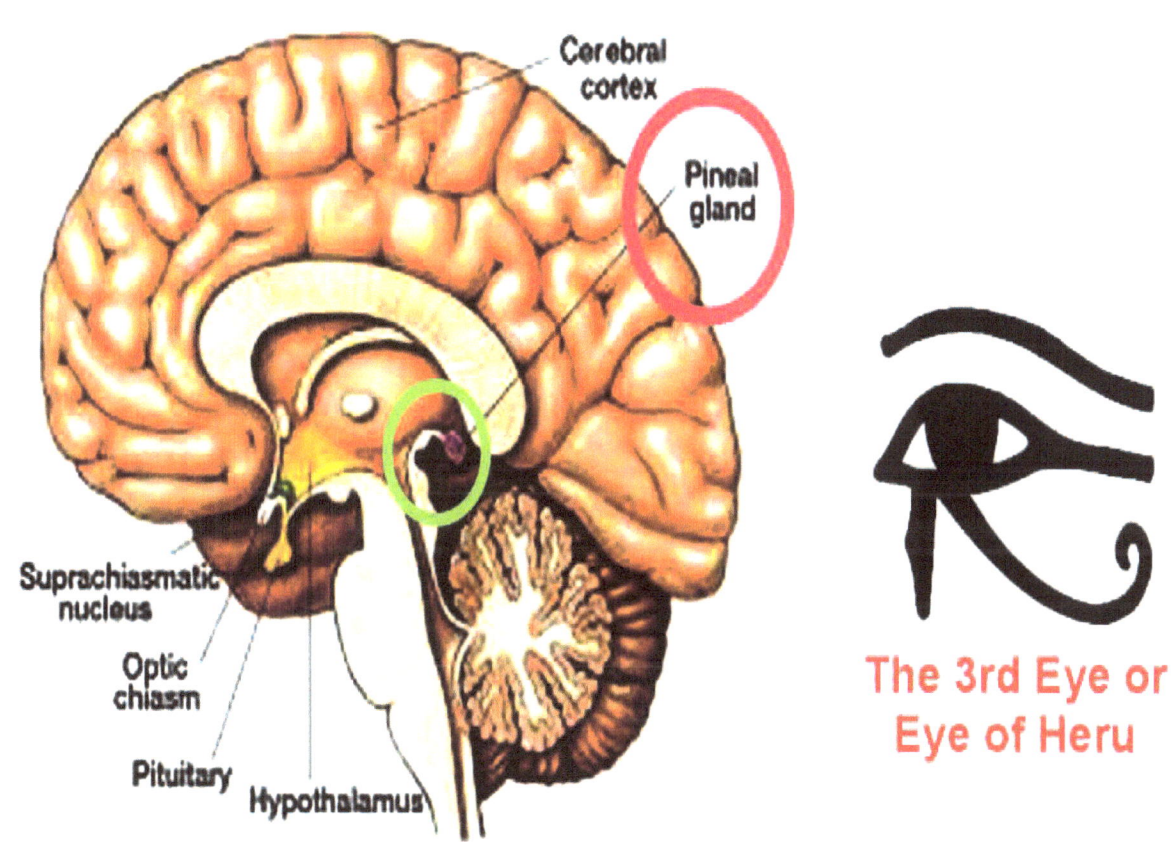

Cerebral cortex

Pineal gland

Suprachiasmatic nucleus

Optic chiasm

Pituitary

Hypothalamus

The 3rd Eye or Eye of Heru

21

How Does Melanin Function?

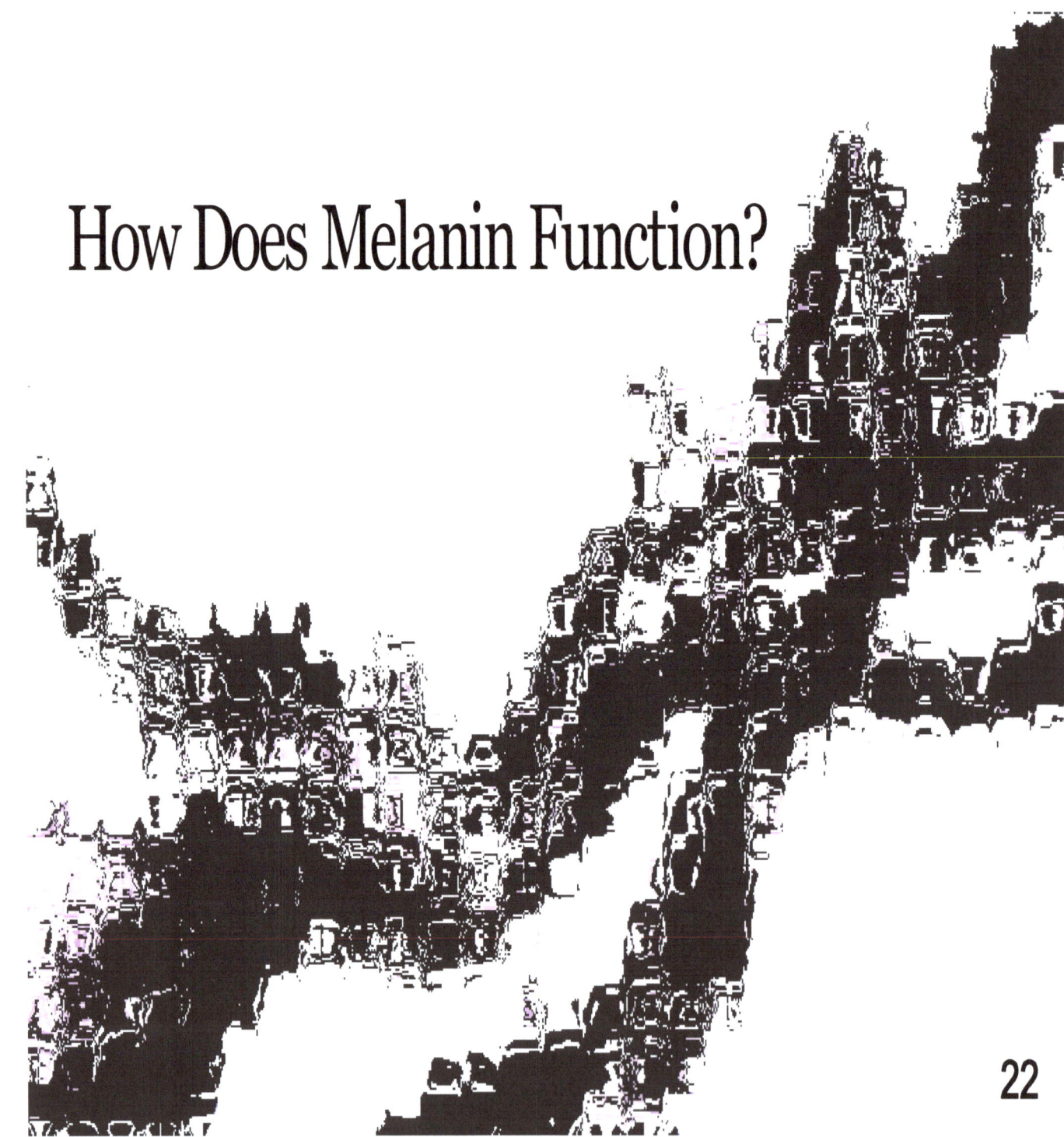

Melanin has several jobs and provides nature with many benefits. It is most known for being a natural sun screen that protects humans from harmful sun rays. In general, those with darker skin and more Melanin can stand exposure to the sun for long periods of time without getting sunburn. On the other hand, someone with less Melanin and little to no pigment has a greater chance of not only being sunburned but also contracting illnesses such as skin cancer. Melanin is also a regulator of all bodily functions and glands. It is a super conductor that assists all organisms in the body in order to control mental and physical activity. Brain Melanin (Neuromelanin) stimulates creativity, strengthens memory and fosters ingenuity. The dark matter called Melanin is a great enhancer to the role of the senses. Due to its amazing properties, it expands the range of sound the ears can hear, makes it possible for the eyes to see colors with intensity and helps the taste buds receive every distinct flavor in foods. A healthy flow of Melanin gives the body the ability to heal itself of all diseases.

24

How do we protect our Melanin?

The presence of Melanin in the human body helps to generate the wonderful flow of life. Getting the proper rest, going to bed before midnight and sleeping six to eight hours, will help to ensure that the Pineal Gland will produce melatonin which makes melanin; also drinking good healthy alkaline water, at least six to eight 8 oz glasses per day, will keep the body hydrated and help support strong Melanin health. Eating raw organic fruits and vegetables, especially those with deep bold colors and leafy greens, will provide the body with great energy that Melanin will use and convert into more great energy. Finally, it is very important that the body moves- walking, dancing and doing other kinds of exercise keeps healthy Melanin flowing in the blood and other cells throughout the entire body. When Melanin health is compromised, it affects the total performance of the human body.

References

Africa, L., 2009. Melanin: What Makes Black People Black. Long Island, NY: Seaburn Publishing Group

King, R. D., 2001. Melanin: A Key to Freedom. Chicago: Lushema Books

Brown, A.C., Bynum, E.B., King, R.D., Moore, T.O., 2005. Why Darkness Matters: The Power of Melanin In the Brain. Chicago: African American Images

Online Resources

www.themelaninimpact.com

www.facebook.com/themelaninimpact

NOTES

www.ingramcontent.com/pod-product-compliance
Lightning Source LLC
Chambersburg PA
CBHW050423180526
45159CB00005B/2392